Differentiated Instruction Teacher Management System

Holt Social Studies

Southwest and Central Asia

HOLT, RINEHART AND WINSTON

A Harcourt Education Company

Orlando • **Austin** • New York • San Diego • Toronto • London

Contents

Teacher Management System

The *Differentiated Instruction Teacher Management System* is designed to assist you in planning and managing instruction for all students who are using *Holt Social Studies: Southwest and Central Asia*. This book has four parts: a Pacing Guide, Section Lesson Plans, Lesson Plans for Differentiated Instruction, and the Teaching Guide and Answer Key for the *Interactive Reader and Study Guide*.

Pacing Guide

This part of the *Differentiated Instruction Teacher Management System* includes a pacing guide—broken down into a 9-week instructional segment and a 12-week instructional segment. The guide suggests materials to be taught and assessed during an average 45-minute class period.

Section Lesson Plans

These on-level lesson plans provide a basic form and planning guide for each section in the textbook. Each lesson plan includes the following content:

- Objectives
- Vocabulary
- Lesson Plan Organization

Lesson Plans for Differentiated Instruction

These lesson plans provide strategies and suggestions for teaching students with special needs. Differentiated strategies are included for:

- English-Language Instruction
- Special Education Instruction
- Advanced/Gifted and Talented Instruction

Teaching Guide and Answer Key for the Interactive Reader and Study Guide

This guide and accompanying answer key are provided to support the *Interactive Reader and Study Guide*, a student workbook that accompanies *Holt Social Studies: Southwest and Central Asia*.

LESSON AND ACTIVITY	9-Week Course	12-Week Course
	(Number of 45-minute class periods)	
Progress Assessment: Diagnostic Test	1	1
Chapter: History of the Fertile Crescent, 7000–500 BC	**8**	**10**
Section 1: Geography of the Fertile Crescent *Geography and History:* River Valley Civilizations	1.5	2.5
Section 2: The Rise of Sumer *Biography:* Sargon *Close-up:* The City-State of Ur	2	2
Section 3: Sumerian Achievements *Connecting to Technology:* The Wheel	1.5	2.5
Section 4: Later Peoples of the Fertile Crescent *Primary Source:* Hammurabi's Code *Social Studies Skills:* Sequencing and Using Time Lines	2	2
Chapter Review and Test	1	1
Chapter: Judaism and Christianity, 2000 BC–AD 1453	**6**	**9**
Section 1: Origins of Judaism *Time Line:* Early Hebrew History *Social Studies Skills:* Interpreting a Route Map	1.5	2.5
Section 2: Origins of Christianity *Focus on Culture:* Christian Holidays	1.5	2.5
Section 3: The Byzantine Empire *Close-up:* The Glory of Constantinople	2	3
Chapter Review and Test	1	1

                    Teacher Management System

Chapter: History of the Islamic World, AD 550–1650	**7.5**	**10**
Section 1: Origins of Islam	1.5	2
Close-up: Life in Arabia		
Time Line: Beginnings of Islam		
Section 2: Islamic Beliefs and Practices	1.5	2
Geography and History: The Hajj		
Section 3: Muslim Empires	1.5	2
Biography: Mehmed II		
Section 4: Cultural Achievements	2	3
Close-up: The Blue Mosque		
Social Studies Skills: Outlining		
Chapter Review and Test	1	1

Chapter: The Eastern Mediterranean	**8**	**10**
Section 1: Physical Geography	1	2
Satellite View: Istanbul and the Bosporus		
Section 2: Turkey	2	2
Close-up: Early Farming Village		
Biography: Kemal Atatürk		
Section 3: Israel	2	2.5
Primary Source: The Dead Sea Scrolls		
World Almanac Facts About Countries: Origin of Israel's Jewish Population		
Focus on Culture: Israeli Teens for Peace		
Social Studies Skills: Analyzing a Cartogram		
Section 4: Syria, Lebanon, and Jordan	2	2.5
Literature: Red Brocade		
Chapter Review and Test	1	1

Chapter: The Arabian Peninsula, Iraq, and Iran	**7.5**	**10**
Section 1: Physical Geography	1	2
Satellite View: Pivot-Irrigated Fields		
Section 2: The Arabian Peninsula	2	3
Connecting to Math: Muslim Contributions to Math		
Case Study: Oil in Saudi Arabia		
Section 3: Iraq	1.5	2
Section 4: Iran	2	2
Biography: Shirin Ebadi		
Social Studies Skills: Analyzing Tables and Statistics		
Chapter Review and Test	1	1

Chapter: Central Asia	**6**	**9**
Section 1: Physical Geography	1.5	2
Section 2: History and Culture	1.5	2.5
Close-up: Inside a Yurt		
Section 3: Central Asia Today	2	3.5
Focus on Culture: Turkmen Carpets		
World Almanac Facts About Countries: Standard of Living in Central Asia		
Geography and History: The Aral Sea		
Social Studies Skills: Using Scale		
Chapter Review and Test	1	1

Progress Assessment: Benchmark Test, Forms A and B	1	1

Additional Teacher Resources to Support Differentiated Instruction
Interactive Reader and Study Guide
Southwest and Central Asia Transparencies
Chapter Visual Summary
Social Studies Skills Activity
Interactive Skills Tutor CD-ROM
Differentiated Instruction Modified Worksheets and Tests CD-ROM

History of the Fertile Crescent

Lesson Plan

Section 1

Objectives *Students will learn that . . .*

1. The rivers of Southwest Asia supported the growth of civilization.

2. New farming techniques led to the growth of cities.

Vocabulary Fertile Crescent, silt, irrigation, canals, surplus, division of labor

PRETEACH	RESOURCES
___ **What You Will Learn . . .** (SE) Preview Main Ideas, Big Idea, and Key Terms. ___ **Key Terms and Places** (TE) Preteach or review the key terms from this section with students. ___ **If YOU lived there . . .** (SE) Have students read and discuss the section introduction and question.	___ Daily Bellringer Transparency: Section 1 ___ RF: Vocabulary Builder: Section 1 ___ Differentiated Instruction Modified Worksheets and Tests CD-ROM
DIRECT TEACH	**RESOURCES**
___ **Teach the Big Idea Activity** (TE) Students discuss the Main Idea questions and then make a proposal for the creation of a memorial to Mesopotamia, commemorating its historical contributions and accomplishments. ___ **Main Idea Questions** (TE) Discussion questions ___ **Map: The Fertile Crescent** (SE) ___ **Critical Thinking: Understanding Cause and Effect** (TE) Cause-and-Effect Posters ___ **Collaborative Learning** (TE) Creating a Farming Community	___ Interactive Reader and Study Guide: Section 1 ___ RF: Geography and History: A Fertile Land ___ Map Zone Transparency: The Fertile Crescent
REVIEW & ASSESS	**RESOURCES**
___ **Close** (TE) Discuss with students the role that water, and the control of it, played in the development of Mesopotamian civilizations. ___ **Section 1 Assessment** (SE)	___ Online Quiz Section 1 (keyword: SGD7 HP1) ___ PASS: Section 1 Quiz

Key: SE = Student Edition **TE** = Teacher's Edition **RF** = Resource File **PASS** = Progress Assessment Support System

History of the Fertile Crescent

Lesson Plan
Section 2

Objectives *Students will learn that . . .*

1. The Sumerians created the world's first advanced society.

2. Religion played a major role in Sumerian society.

Vocabulary Sumer, city-state, empire, polytheism, priests, social hierarchy

PRETEACH	RESOURCES
___ **What You Will Learn . . .** (SE) Preview Main Ideas, Big Idea, and Key Terms. ___ **Academic Vocabulary** (TE) Review with students the high-use academic terms in this section. ___ **If YOU lived there . . .** (SE) Have students read and discuss the section introduction and question.	___ Daily Bellringer Transparency: Section 2 ___ RF: Vocabulary Builder: Section 2 ___ Differentiated Instruction Modified Worksheets and Tests CD-ROM

DIRECT TEACH	RESOURCES
___ **Teach the Big Idea Activity** (TE) Students discuss the Main Idea questions and then create a chart examining the headings, subheadings, and bold terms from this section. ___ **Main Idea Questions** (TE) Discussion questions ___ **Map: Sargon's Empire, c. 2330 BC** (SE) ___ **Biography: Sargon** (SE) ___ **Cross-Discipline Activity: Literature** (TE) Writing an Autobiography ___ **Close-up: The City-State of Ur** (SE) ___ **Differentiating Instruction: Struggling Readers** (TE) Rise of Sumer Graphic Organizer ___ **Critical Thinking: Summarizing** (TE) Illustrated Social Hierarchy	___ Interactive Reader and Study Guide: Section 2 ___ Map Zone Transparency: Sargon's Empire, c. 2330 BC ___ RF: Primary Source: The Sumerian Flood Story ___ RF: Biography: Enheduanna

REVIEW & ASSESS	RESOURCES
___ **Close** (TE) Have students write a short paragraph summarizing the government, religion, and society of Sumer. ___ **Section 2 Assessment** (SE)	___ Online Quiz Section 2 (keyword: SGD7 HP1) ___ PASS: Section 2 Quiz

Key: SE = Student Edition **TE** = Teacher's Edition **RF** = Resource File **PASS** = Progress Assessment Support System

   Teacher Management System

History of the Fertile Crescent

Lesson Plan
Section 3

Objectives *Students will learn that . . .*

1. The Sumerians invented the world's first writing system.

2. Advances and inventions changed Sumerian lives.

3. Many types of art developed in Sumer.

Vocabulary cuneiform, pictographs, scribe, epics, architecture, ziggurat

PRETEACH	RESOURCES
___ **What You Will Learn . . .** (SE) Preview Main Ideas, Big Idea, and Key Terms. ___ **Academic Vocabulary** (TE) Review with students the high-use academic term in this section. ___ **If YOU lived there . . .** (SE) Have students read and discuss the section introduction and question.	___ Daily Bellringer Transparency: Section 3 ___ RF: Vocabulary Builder: Section 3 ___ Differentiated Instruction Modified Worksheets and Tests CD-ROM
DIRECT TEACH	**RESOURCES**
___ **Teach the Big Idea Activity** (TE) Students discuss the Main Idea questions and then identify different Sumerian achievements, determining which, if any, continue to impact the world today. ___ **Main Idea Questions** (TE) Discussion questions ___ **Differentiating Instruction: English-Language Learners** (TE) Pictograph Activity ___ **Connecting to Technology: The Wheel** (SE) ___ **Differentiating Instruction: Special Needs Learners** (TE) Sumerian Inventions ___ **Differentiating Instruction: Advanced/Gifted and Talented** (TE) Sumerian Numerals ___ **Collaborative Learning** (TE) Creating a Television Commercial	___ Interactive Reader and Study Guide: Section 3 ___ RF: Interdisciplinary Project: Mesopotamia: The First Writing ___ RF: Literature: The Epic of Gilgamesh ___ RF: Interdisciplinary Project: From Wheel to Gear: Making a Model
REVIEW & ASSESS	**RESOURCES**
___ **Close** (TE) Ask students to name Sumerian achievements and describe why they were important. ___ **Section 3 Assessment** (SE)	___ Online Quiz Section 3 (keyword: SGD7 HP1) ___ PASS: Section 3 Quiz

Key: SE = Student Edition **TE** = Teacher's Edition **RF** = Resource File **PASS** = Progress Assessment Support System

                                    Teacher Management System

History of the Fertile Crescent

Lesson Plan

Section 4

Objectives *Students will learn that . . .*

1. The Babylonians conquered Mesopotamia and created a code of law.

2. Invasions of Mesopotamia changed the region's culture.

3. The Phoenicians built a trading society in the eastern Mediterranean region.

Vocabulary Babylon, Hammurabi's Code, chariot, alphabet

PRETEACH	RESOURCES
___ **What You Will Learn . . .** (SE) Preview Main Ideas, Big Idea, and Key Terms. ___ **Key Terms and Places** (TE) Preteach or review the key terms from this section with students. ___ **If YOU lived there . . .** (SE) Have students read and discuss the section introduction and question.	___ Daily Bellringer Transparency: Section 4 ___ RF: Vocabulary Builder: Section 4 ___ Differentiated Instruction Modified Worksheets and Tests CD-ROM
DIRECT TEACH	**RESOURCES**
___ **Teach the Big Idea Activity** (TE) Students discuss the Main Idea questions and then create a time line that includes the later empires and kingdoms that developed in Mesopotamia. ___ **Main Idea Questions** (TE) Discussion questions ___ **Primary Source: Hammurabi's Code** (SE) ___ **Differentiating Instruction: Advanced/Gifted and Talented** (TE) A World Without Written Laws ___ **Map: Babylonian and Assyrian Empires** (SE) ___ **Differentiating Instruction: Struggling Readers** (TE) Hittite Letter ___ **Critical Thinking: Summarizing** (TE) Creating a Book Jacket ___ **Map: Phoenicia, c. 800 BC** (SE) ___ **Critical Thinking: Analyzing** (TE) Phoenician Exports ___ **Social Studies Skills: Sequencing and Using Time Lines** (SE)	___ Interactive Reader and Study Guide: Section 4 ___ RF: Biography: Hammurabi ___ RF: Primary Source: The Code of Hammurabi ___ Map Zone Transparency: Babylonian and Assyrian Empires ___ RF: Biography: King Nebuchadnezzar ___ RF: Primary Source: Descriptions of the Phoenicians ___ Map Zone Transparency: Phoenicia, c. 800 BC
REVIEW & ASSESS	**RESOURCES**
___ **Close** (TE) Discuss with students the order of peoples that ruled Mesopotamia and how each group came to power. ___ **Section 4 Assessment** (SE)	___ Quick Facts Transparency: History of the Fertile Crescent Visual Summary ___ Online Quiz Section 4 (keyword: SGD7 HP1) ___ PASS: Section 4 Quiz

Key: SE = Student Edition **TE** = Teacher's Edition **RF** = Resource File **PASS** = Progress Assessment Support System

  Teacher Management System

Judaism and Christianity

Lesson Plan

Section 1

Objectives *Students will learn that . . .*

1. The Hebrews' early history began in Canaan and ended when the Romans forced them out of Israel.

2. Jewish beliefs in God, justice, and law anchor their society.

3. Jewish sacred texts describe the laws and principles of Judaism.

4. Traditions and holy days celebrate the history and religion of the Jewish people.

Vocabulary Judaism, Canaan, Exodus, monotheism, Torah, rabbis

PRETEACH	RESOURCES
___ **What You Will Learn . . .** (SE) Preview Main Ideas, Big Idea, and Key Terms.	___ Daily Bellringer Transparency: Section 1
___ **Academic Vocabulary** (TE) Review with students the high-use academic term in this section.	___ RF: Vocabulary Builder: Section 1
___ **If YOU lived there . . .** (SE) Have students read and discuss the section introduction and question.	___ Differentiated Instruction Modified Worksheets and Tests CD-ROM

DIRECT TEACH	RESOURCES
___ **Teach the Big Idea Activity** (TE) Students discuss the Main Idea questions and then examine the origins of Judaism.	___ Interactive Reader and Study Guide: Section 1
___ **Main Idea Questions** (TE) Discussion questions	___ Map Zone Transparency: Jewish Migration after AD 70
___ **Time Line: Early Hebrew History** (SE)	
___ **Critical Thinking: Evaluating Information** (TE) Writing Headlines	___ RF: Biography: David
___ **Map: Jewish Migration after AD 70** (SE)	___ RF: Literature: *The Chosen*
___ **Differentiating Instruction: Struggling Readers** (TE) Creating a Jewish Settlement Chart	
___ **Critical Thinking: Comparing and Contrasting** (TE) Analyzing Jewish Sects	
___ **Differentiating Instruction: English-Language Learners** (TE) Linking Visuals to Text	
___ **Differentiating Instruction: Advanced/Gifted and Talented** (TE) Researching Proverbs from the Hebrew Bible	
___ **Collaborative Learning** (TE) Exploring Different Languages	
___ **Social Studies Skills: Interpreting a Route Map** (SE)	

REVIEW & ASSESS	RESOURCES
___ **Close** (TE) Use the section's vocabulary to review important information about early Hebrew history and the practice of Judaism today.	___ Online Quiz Section 1 (keyword: SGD7 HP2)
___ **Section 1 Assessment** (SE)	___ PASS: Section 1 Quiz

Key: SE = Student Edition **TE** = Teacher's Edition **RF** = Resource File **PASS** = Progress Assessment Support System

　　　　　　　　　　　　　Teacher Management System

Judaism and Christianity

Lesson Plan

Section 2

Objectives *Students will learn that . . .*

1. The life and death of Jesus of Nazareth inspired a new religion called Christianity.

2. Christians believe that Jesus's acts and teachings focused on love and salvation.

3. Jesus's followers taught others about Jesus's life and teachings.

4. Christianity spread throughout the Roman Empire by 400.

Vocabulary Messiah, Christianity, Bible, Bethlehem, Resurrection, disciples, saint

PRETEACH	RESOURCES
___ **What You Will Learn . . .** (SE) Preview Main Ideas, Big Idea, and Key Terms. ___ **Academic Vocabulary** (TE) Review with students the high-use academic term in this section. ___ **If YOU lived there . . .** (SE) Have students read and discuss the section introduction and question.	___ Daily Bellringer Transparency: Section 2 ___ RF: Vocabulary Builder: Section 2 ___ Differentiated Instruction Modified Worksheets and Tests CD-ROM

DIRECT TEACH	RESOURCES
___ **Teach the Big Idea Activity** (TE) Students discuss the Main Idea questions and then make a sequence chart showing the events that led to the spread of Christianity. ___ **Main Idea Questions** (TE) Discussion questions ___ **Differentiating Instruction: English-Language Learners** (TE) Multiple Word Meanings ___ **Focus on Culture: Christian Holidays** (SE) ___ **Differentiating Instruction: Struggling Readers** (TE) Completing a Cluster Diagram ___ **Critical Thinking: Analyzing Information** (TE) Creating a Briefing Paper ___ **Critical Thinking: Drawing Conclusions** (TE) Creating a Web Log ___ **Map: Paul's Journeys** (SE) ___ **Cross-Discipline Activity: English/Language Arts** (TE) Writing a Narrative Poem ___ **Map: The Spread of Christianity, 300–400** (SE) ___ **Critical Thinking: Analyzing Primary Sources** (TE) Analyzing the Edict of Milan	___ Interactive Reader and Study Guide: Section 2 ___ RF: Critical Thinking: Masada ___ RF: Biography: Kathleen Kenyon ___ Map Zone Transparency: Paul's Journeys ___ RF: Geography For Life: The Spread of Christianity ___ RF: Primary Source: The Edict of Milan ___ Map Zone Transparency: The Spread of Christianity, 300–400

REVIEW & ASSESS	RESOURCES
___ **Close** (TE) Review the sequence of events and developments that led to the beginning, growth, and spread of Christianity. ___ **Section 2 Assessment** (SE)	___ Online Quiz Section 2 (keyword: SGD7 HP2) ___ PASS: Section 2 Quiz

Key: SE = Student Edition **TE** = Teacher's Edition **RF** = Resource File **PASS** = Progress Assessment Support System

Judaism and Christianity

Lesson Plan

Section 3

Objectives *Students will learn that . . .*

1. Eastern emperors ruled from Constantinople and tried but failed to reunite the whole Roman Empire.

2. The people of the eastern empire created a new society that was very different from society in the west.

3. Byzantine Christianity was different from religion in the west.

Vocabulary Constantinople, Byzantine Empire, mosaics

PRETEACH	RESOURCES
___ **What You Will Learn . . .** (SE) Preview Main Ideas, Big Idea, and Key Terms. ___ **Key Terms and Places** (TE) Preteach or review the key terms from this section with students. ___ **If YOU lived there . . .** (SE) Have students read and discuss the section introduction and question.	___ Daily Bellringer Transparency: Section 3 ___ RF: Vocabulary Builder: Section 3 ___ Differentiated Instruction Modified Worksheets and Tests CD-ROM

DIRECT TEACH	RESOURCES
___ **Teach the Big Idea Activity** (TE) Students discuss the Main Idea questions and then create a chart on Byzantine emperors and Byzantine culture. ___ **Main Idea Questions** (TE) Discussion questions ___ **Map: The Byzantine Empire, 1025** (SE) ___ **Critical Thinking: Analyzing Information** (TE) Creating a Time Line ___ **Close-up: The Glory of Constantinople** (SE) ___ **Differentiating Instruction: Advanced/Gifted and Talented** (TE) Writing a Tourist Brochure ___ **Differentiating Instruction: Special Needs Learners** (TE) Discussing Parades and Processions ___ **Cross-Discipline Activity: Art** (TE) Creating Mosaics	___ Interactive Reader and Study Guide: Section 3 ___ Map Zone Transparency: The Byzantine Empire, 1025 ___ RF: Geography and History: The First Crusade

REVIEW & ASSESS	RESOURCES
___ **Close** (TE) Briefly review the differences between the eastern and western empires. ___ **Section 3 Assessment** (SE)	___ Quick Facts Transparency: Judaism and Christianity Visual Summary ___ Online Quiz Section 3 (keyword: SGD7 HP2) ___ PASS: Section 3 Quiz

Key: SE = Student Edition **TE** = Teacher's Edition **RF** = Resource File **PASS** = Progress Assessment Support System

History of the Islamic World

Lesson Plan

Section 1

Objectives *Students will learn that . . .*

1. Arabia is mostly a desert land, where two ways of life, nomadic and sedentary, developed.

2. A new religion called Islam, founded by the prophet Muhammad, spread throughout Arabia in the 600s.

Vocabulary Mecca, Islam, Muslim, Qur´an, Medina, mosque

PRETEACH	RESOURCES
___ **What You Will Learn . . .** (SE) Preview Main Ideas, Big Idea, and Key Terms. ___ **Key Terms and Places** (TE) Preteach or review the key terms from this section with students. ___ **If YOU lived there . . .** (SE) Have students read and discuss the section introduction and question.	___ Daily Bellringer Transparency: Section 1 ___ RF: Vocabulary Builder: Section 1 ___ Differentiated Instruction Modified Worksheets and Tests CD-ROM

DIRECT TEACH	RESOURCES
___ **Teach the Big Idea Activity** (TE) Students discuss the Main Idea questions and then create their own main idea statements. ___ **Main Idea Questions** (TE) Discussion questions ___ **Close-up: Life in Arabia** (SE) ___ **Critical Thinking: Making Generalizations** (TE) A Day in the Life ___ **Time Line: Beginnings of Islam** (SE) ___ **Differentiating Instruction: Advanced/Gifted and Talented** (TE) Christianity, Judaism, and Islam	___ Interactive Reader and Study Guide: Section 1 ___ RF: Biography: Khadijah ___ RF: Primary Source: Reading *from* The Qur´an

REVIEW & ASSESS	RESOURCES
___ **Close** (TE) Have students work in pairs to write one-paragraph summaries of the geography of Arabia. Then have students discuss the origins of Islam as well as early reactions to Muhammad and his teachings. ___ **Section 1 Assessment** (SE)	___ Online Quiz Section 1 (keyword: SGD7 HP3) ___ PASS: Section 1 Quiz

Key: SE = Student Edition **TE** = Teacher's Edition **RF** = Resource File **PASS** = Progress Assessment Support System

History of the Islamic World

Lesson Plan

Section 2

Objectives *Students will learn that . . .*

1. The Qur´an guides Muslims' lives.

2. The Sunnah tells Muslims of important duties expected of them.

3. Islamic law is based on the Qur´an and the Sunnah.

Vocabulary jihad, Sunnah, Five Pillars of Islam

PRETEACH	RESOURCES
___ **What You Will Learn . . .** (SE) Preview Main Ideas, Big Idea, and Key Terms. ___ **Academic Vocabulary** (TE) Review with students the high-use academic terms in this section. ___ **If YOU lived there . . .** (SE) Have students read and discuss the section introduction and question.	___ Daily Bellringer Transparency: Section 2 ___ RF: Vocabulary Builder: Section 2 ___ Differentiated Instruction Modified Worksheets and Tests CD-ROM
DIRECT TEACH	**RESOURCES**
___ **Teach the Big Idea Activity** (TE) Students discuss the Main Idea questions and then create a poster that lists important beliefs, guidelines, and practices of Islam. ___ **Main Idea Questions** (TE) Discussion questions ___ **Differentiating Instruction: Special Needs Learners** (TE) The Qur´an ___ **Critical Thinking: Summarizing** (TE) The Five Pillars of Islam Book Jacket	___ Interactive Reader and Study Guide: Section 2
REVIEW & ASSESS	**RESOURCES**
___ **Close** (TE) Review the three sources of Islamic beliefs with students. ___ **Section 2 Assessment** (SE)	___ Online Quiz Section 2 (keyword: SGD7 HP3) ___ PASS: Section 2 Quiz

Key: SE = Student Edition **TE** = Teacher's Edition **RF** = Resource File **PASS** = Progress Assessment Support System

  Teacher Management System

History of the Islamic World

Lesson Plan
Section 3

Objectives *Students will learn that . . .*

1. Muslim armies conquered many lands into which Islam slowly spread.
2. Trade helped Islam spread into new areas.
3. Three Muslim empires controlled much of Europe, Asia, and Africa from the 1400s to the 1800s.

Vocabulary caliph, tolerance, Baghdad, Córdoba, janissaries, Istanbul, Esfahan

PRETEACH	RESOURCES
___ **What You Will Learn . . .** (SE) Preview Main Ideas, Big Idea, and Key Terms. ___ **Key Terms and Places** (TE) Preteach or review the key terms from this section with students. ___ **If YOU lived there . . .** (SE) Have students read and discuss the section introduction and question.	___ Daily Bellringer Transparency: Section 3 ___ RF: Vocabulary Builder: Section 3 ___ Differentiated Instruction Modified Worksheets and Tests CD-ROM

DIRECT TEACH	RESOURCES
___ **Teach the Big Idea Activity** (TE) Students discuss the Main Idea questions and then identify cause and effect in the spread of Islam after Muhammad's death. ___ **Main Idea Questions** (TE) Discussion questions ___ **Differentiating Instruction: Advanced/Gifted and Talented** (TE) Muslim Architecture ___ **Critical Thinking: Comparing and Contrasting** (TE) Explaining Trading Goods and Ideas ___ **Map: The Ottoman Empire** (SE) ___ **Biography: Mehmed II** (SE) ___ **Differentiating Instruction: Advanced/Gifted and Talented** (TE) The Ottoman Conquest ___ **Map: The Safavid Empire** (SE) ___ **Differentiating Instruction: English-Language Learners** (TE) Culture of the Safavid Empire ___ **Map: The Mughal Empire** (SE)	___ Interactive Reader and Study Guide: Section 3 ___ RF: Geography and History: Arabia, AD 750 ___ RF: Primary Source: Jeweled Canteen ___ Map Zone Transparency: The Ottoman Empire ___ RF: Biography: Akbar the Great ___ RF: Biography: Mumtaz Mahal ___ Map Zone Transparency: The Safavid Empire

REVIEW & ASSESS	RESOURCES
___ **Close** (TE) Briefly review the ways in which Islam spread throughout the Ottoman, Safavid, and Mughal empires. Point out that by the end of the 1600s, Islam was the dominant religion in the region. ___ **Section 3 Assessment** (SE)	___ Online Quiz Section 3 (keyword: SGD7 HP3) ___ PASS: Section 3 Quiz

Key: **SE** = Student Edition **TE** = Teacher's Edition **RF** = Resource File **PASS** = Progress Assessment Support System

History of the Islamic World

Lesson Plan

Section 4

Objectives *Students will learn that . . .*

1. Muslim scholars made lasting contributions to the fields of science and philosophy.

2. In literature and the arts, Muslim achievements included beautiful poetry, memorable short stories, and splendid architecture.

Vocabulary Sufism, minarets, calligraphy

PRETEACH	RESOURCES
___ **What You Will Learn . . .** (SE) Preview Main Ideas, Big Idea, and Key Terms. ___ **Academic Vocabulary** (TE) Review with students the high-use academic term in this section. ___ **If YOU lived there . . .** (SE) Have students read and discuss the section introduction and question.	___ Daily Bellringer Transparency: Section 4 ___ RF: Vocabulary Builder: Section 4 ___ Differentiated Instruction Modified Worksheets and Tests CD-ROM
DIRECT TEACH	**RESOURCES**
___ **Teach the Big Idea Activity** (TE) Students discuss the Main Idea questions and then take notes on an assigned figure or cultural achievement. ___ **Main Idea Questions** (TE) Discussion questions ___ **Critical Thinking: Finding Main Ideas** (TE) Muslim Achievements: "New and Improved" Ads ___ **Close-up: The Blue Mosque** (SE) ___ **Cross-Discipline Activity: Arts and the Humanities** (TE) Mosque Web Site ___ **Social Studies Skills: Outlining** (SE)	___ Interactive Reader and Study Guide: Section 4 ___ RF: Primary Source: Jahangir Presenting a Book to a Sufi
REVIEW & ASSESS	**RESOURCES**
___ **Close** (TE) Ask students how their lives might be different now because of the achievements of the Islamic empires. Have students consider all the various Islamic achievements, from mathematics to literature. ___ **Section 4 Assessment** (SE)	___ Quick Facts Transparency: History of the Islamic World Visual Summary ___ Online Quiz Section 4 (keyword: SGD3 HP3) ___ PASS: Section 4 Quiz

Key: SE = Student Edition **TE** = Teacher's Edition **RF** = Resource File **PASS** = Progress Assessment Support System

The Eastern Mediterranean

Lesson Plan
Section 1

Objectives *Students will learn that . . .*

1. The Eastern Mediterranean's physical features include the Bosporus, the Dead Sea, rivers, mountains, deserts, and plains.

2. The region's climate is mostly dry with little vegetation.

3. Important natural resources in the Eastern Mediterranean include valuable minerals and the availability of water.

Vocabulary Dardanelles, Bosporus, Jordan River, Dead Sea, Syrian Desert

PRETEACH	RESOURCES
___ **What You Will Learn . . .** (SE) Preview Main Ideas, Big Idea, and Key Terms. ___ **Key Terms and Places** (TE) Preteach or review the key terms from this section with students. ___ **If YOU lived there . . .** (SE) Have students read and discuss the section introduction and question.	___ Daily Bellringer Transparency: Section 1 ___ RF: Vocabulary Builder: Section 1 ___ Differentiated Instruction Modified Worksheets and Tests CD-ROM
DIRECT TEACH	**RESOURCES**
___ **Teach the Big Idea Activity** (TE) Students discuss the Main Idea questions and then choose a physical feature found in the Eastern Mediterranean that is different from any features in the region they live in. ___ **Main Idea Questions** (TE) Discussion questions ___ **Map: The Eastern Mediterranean: Physical** (SE) ___ **Differentiating Instruction: Special Needs Learners** (TE) Learning From Visuals ___ **Differentiating Instruction: Advanced/Gifted and Talented** (TE) Linking to Today ___ **Satellite View: Istanbul and the Bosporus** (SE) ___ **Collaborative Learning** (TE) Climate Maps ___ **Map: The Eastern Mediterranean: Climate** (SE)	___ Interactive Reader and Study Guide: Section 1 ___ Map Zone Transparency: The Eastern Mediterranean: Physical ___ Map Zone Transparency: The Eastern Mediterranean: Climate
REVIEW & ASSESS	**RESOURCES**
___ **Close** (TE) Discuss with students important aspects of the physical features, climate, vegetation, and natural resources that play a role in the Eastern Mediterranean. ___ **Section 1 Assessment** (SE)	___ Online Quiz Section 1 (keyword: SGD7 HP4) ___ PASS: Section 1 Quiz

Key: SE = Student Edition **TE** = Teacher's Edition **RF** = Resource File **PASS** = Progress Assessment Support System

 Teacher Management System

The Eastern Mediterranean

Lesson Plan
Section 2

Objectives *Students will learn that . . .*

1. Turkey's history includes invasion by the Romans, rule by the Ottomans, and a twentieth-century democracy.

2. Turkey's people are mostly ethnic Turks, and its culture is a mixture of modern and traditional.

3. Today, Turkey is a democratic nation seeking economic opportunities as a future member of the European Union.

Vocabulary Ankara, Istanbul, secular

PRETEACH	RESOURCES
___ **What You Will Learn . . .** (SE) Preview Main Ideas, Big Idea, and Key Terms.	___ Daily Bellringer Transparency: Section 2
___ **Academic Vocabulary** (TE) Review with students the high-use academic term in this section.	___ RF: Vocabulary Builder: Section 2
___ **If YOU lived there . . .** (SE) Have students read and discuss the section introduction and question.	___ Differentiated Instruction Modified Worksheets and Tests CD-ROM

DIRECT TEACH	RESOURCES
___ **Teach the Big Idea Activity** (TE) Students discuss the Main Idea questions and then create a chart of invasions throughout Turkey's history.	___ Interactive Reader and Study Guide: Section 2
___ **Main Idea Questions** (TE) Discussion questions	___ RF: Geography and History: Growth of the Ottoman Empire
___ **Close-up: Early Farming Village** (SE)	
___ **Biography: Kemal Atatürk** (SE)	___ RF: Biography: Suleyman I
___ **Cross-Discipline Activity: English/Language Arts** (TE) Writing an Autobiography	___ Map Zone Transparency: Turkey: Population
___ **Map: Turkey: Population** (SE)	___ RF: Primary Source: Atatürk's Address to Turkish Youth
___ **Collaborative Learning** (TE) More about Turkey	___ RF: Critical Thinking: The Adventures of Mulla Nasrudin

REVIEW & ASSESS	RESOURCES
___ **Close** (TE) Briefly review Turkey's history, people, government, and economy.	___ Online Quiz Section 2 (keyword: SGD7 HP4)
___ **Section 2 Assessment** (SE)	___ PASS: Section 2 Quiz

Key: SE = Student Edition **TE** = Teacher's Edition **RF** = Resource File **PASS** = Progress Assessment Support System

                                    Teacher Management System

The Eastern Mediterranean

Lesson Plan

Section 3

Objectives *Students will learn that . . .*

1. Israel's history includes the ancient Hebrews and the creation of the nation of Israel.

2. In Israel today, Jewish culture is a major part of daily life.

3. The Palestinian Territories are areas within Israel controlled partly by Palestinian Arabs.

Vocabulary Diaspora, Jerusalem, Zionism, kosher, kibbutz, Gaza, West Bank

PRETEACH	RESOURCES
___ **What You Will Learn . . .** (SE) Preview Main Ideas, Big Idea, and Key Terms. ___ **Key Terms and Places** (TE) Preteach or review the key terms from this section with students. ___ **If YOU lived there . . .** (SE) Have students read and discuss the section introduction and question.	___ Daily Bellringer Transparency: Section 3 ___ RF: Vocabulary Builder: Section 3 ___ Differentiated Instruction Modified Worksheets and Tests CD-ROM

DIRECT TEACH	RESOURCES
___ **Teach the Big Idea Activity** (TE) Students discuss the Main Idea questions and then create a chart with facts about Israel and the Palestinian Territories. ___ **Main Idea Questions** (TE) Discussion questions ___ **Primary Source Activity: The Dead Sea Scrolls** (SE) ___ **Differentiating Instruction: English-Language Learners** (TE) History of Israel Time Line ___ **World Almanac: Facts about Countries: Origin of Israel's Jewish Population** (SE) ___ **Collaborative Learning** (TE) Life on a Kibbutz ___ **Map: Israel and the Palestinian Territories** (SE) ___ **Critical Thinking: Analyzing Information** (TE) History in the Making ___ **Focus on Culture: Israeli Teens for Peace** (SE) ___ **Social Studies Skills: Analyzing a Cartogram** (SE)	___ Interactive Reader and Study Guide: Section 3 ___ RF: Biography: Golda Meir ___ Map Zone Transparency: Israel and the Palestinian Territories ___ RF: Geography for Life: What Future for Jerusalem? ___ RF: Interdisciplinary Project, Civics and Government: News Report on the Palestinian-Israeli Conflict

REVIEW & ASSESS	RESOURCES
___ **Close** (TE) Briefly review Israel's history, people, government, economy, and the Palestinian Territories. ___ **Section 3 Assessment** (SE)	___ Online Quiz Section 3 (keyword: SGD7 HP4) ___ PASS: Section 3 Quiz

Key: SE = Student Edition **TE** = Teacher's Edition **RF** = Resource File **PASS** = Progress Assessment Support System

The Eastern Mediterranean

Lesson Plan

Section 4

Objectives *Students will learn that . . .*

1. Syria, once part of the Ottoman Empire, is an Arab country ruled by a powerful family.

2. Lebanon is recovering from civil war and its people are divided by religion.

3. Jordan has few resources and is home to Bedouins and Palestinian refugees.

Vocabulary Damascus, Beirut, Bedouins, Amman

PRETEACH	RESOURCES
___ **What You Will Learn . . .** (SE) Preview Main Ideas, Big Idea, and Key Terms. ___ **Key Terms and Places** (TE) Preteach or review the key terms from this section with students ___ **If YOU lived there . . .** (SE) Have students read and discuss the section introduction and question.	___ Daily Bellringer Transparency: Section 4 ___ RF: Vocabulary Builder: Section 4 ___ Differentiated Instruction Modified Worksheets and Tests CD-ROM
DIRECT TEACH	**RESOURCES**
___ **Teach the Big Idea Activity** (TE) Students discuss the Main Idea questions and then work in groups to create charts on Syria, Lebanon, and Jordan. ___ **Main Idea Questions** (TE) Discussion questions ___ **Critical Thinking: Summarizing** (TE) The Region's Main Religions ___ **Critical Thinking: Compare and Contrast** (TE) Syria, Lebanon, and Jordan	___ Interactive Reader and Study Guide: Section 4 ___ RF: Literature: "Friendship"
REVIEW & ASSESS	**RESOURCES**
___ **Close** (TE) Briefly review the similarities and differences of Syria, Lebanon, and Jordan. ___ **Section 4 Assessment** (SE)	___ Quick Facts Transparency: The Eastern Mediterranean Visual Summary ___ Online Quiz Section 4 (keyword: SGD7 HP4) ___ PASS: Section 4 Quiz

Key: **SE** = Student Edition **TE** = Teacher's Edition **RF** = Resource File **PASS** = Progress Assessment Support System

The Arabian Peninsula, Iraq, and Iran

Lesson Plan

Section 1

Objectives *Students will learn that . . .*

1. Major physical features of the Arabian Peninsula, Iraq, and Iran are desert plains and mountains.

2. The region has a dry climate and little vegetation.

3. Most of the world is dependent on oil, a resource that is exported from this region.

Vocabulary Arabian Peninsula, Persian Gulf, Tigris River, Euphrates River, oasis, wadis, fossil water

PRETEACH	RESOURCES
___ **What You Will Learn . . .** (SE) Preview Main Ideas, Big Idea, and Key Terms.	___ Daily Bellringer Transparency: Section 1
___ **Key Terms and Places** (TE) Preteach or review the key terms from this section with students.	___ RF: Vocabulary Builder: Section 1
___ **If YOU lived there . . .** (SE) Have students read and discuss the section introduction and question.	___ Differentiated Instruction Modified Worksheets and Tests CD-ROM

DIRECT TEACH	RESOURCES
___ **Teach the Big Idea Activity** (TE) Students discuss the Main Idea questions and then create a map of the region.	___ Interactive Reader and Study Guide: Section 1
___ **Main Idea Questions** (TE) Discussion questions	___ Map Zone Transparency: Arabian Peninsula, Iraq, and Iran: Physical
___ **Map: Arabian Peninsula, Iraq, and Iran: Physical** (SE)	
___ **Critical Thinking: Summarizing** (TE) Bodies of Water	___ RF: Literature: The Story of Sinbad the Sailor
___ **Map: The Arabian Peninsula, Iraq, and Iran: Climate** (SE)	
___ **Differentiating Instruction: English-Language Learners** (TE) Climate and Vegetation	___ RF: Interdisciplinary Project: *The Arabian Nights:* Write a Story
___ **Differentiating Instruction: Struggling Readers** (TE) Map Activity	
___ **Satellite View: Pivot-Irrigated Fields** (SE)	___ Map Zone Transparency: The Arabian Peninsula, Iraq, and Iran: Climate

REVIEW & ASSESS	RESOURCES
___ **Close** (TE) Ask students to identify the main features of this region's physical geography.	___ Online Quiz Section 1 (keyword: SGD7 HP5)
___ **Section 1 Assessment** (SE)	___ PASS: Section 1 Quiz

Key: SE = Student Edition **TE** = Teacher's Edition **RF** = Resource File **PASS** = Progress Assessment Support System

The Arabian Peninsula, Iraq, and Iran

Lesson Plan

Section 2

Objectives *Students will learn that . . .*

1. Islamic culture and an economy greatly based on oil influence life in Saudi Arabia.

2. Most other Arabian Peninsula countries are monarchies influenced by Islamic culture and oil resources.

Vocabulary Shia, Sunni, OPEC

PRETEACH	RESOURCES
___ **What You Will Learn . . .** (SE) Preview Main Ideas, Big Idea, and Key Terms. ___ **Academic Vocabulary** (TE) Review with students the high-use academic term in this section. ___ **If YOU lived there . . .** (SE) Have students read and discuss the section introduction and question.	___ Daily Bellringer Transparency: Section 2 ___ RF: Vocabulary Builder: Section 2 ___ Differentiated Instruction Modified Worksheets and Tests CD-ROM
DIRECT TEACH	**RESOURCES**
___ **Teach the Big Idea Activity** (TE) Students discuss the Main Idea questions and then create a scrapbook about the Arabian Peninsula. ___ **Main Idea Questions** (TE) Discussion questions ___ **Connecting to Math: Muslim Contributions to Math** (SE) ___ **Collaborative Learning** (TE) Saudi Arabia Test ___ **Differentiating Instruction: Special Needs Learners** (TE) Matching Game	___ Interactive Reader and Study Guide: Section 2 ___ RF: Biography: Ibn Saud ___ RF: Geography For Life: The Hajj
REVIEW & ASSESS	**RESOURCES**
___ **Close** (TE) Ask students to explain the importance of Islam, monarchy, and oil to the Arabian Peninsula. ___ **Section 2 Assessment** (SE)	___ Online Quiz Section 2 (keyword: SGD7 HP5) ___ PASS: Section 2 Quiz

Key: SE = Student Edition **TE** = Teacher's Edition **RF** = Resource File **PASS** = Progress Assessment Support System

The Arabian Peninsula, Iraq, and Iran

Lesson Plan

Section 3

Objectives *Students will learn that . . .*

1. Iraq's history includes rule by many conquerors and cultures, as well as recent wars.

2. Most of Iraq's people are Arabs, and Iraqi culture includes the religion of Islam.

3. Iraq today must rebuild its government and economy, which have suffered from years of conflict.

Vocabulary embargo, Baghdad

PRETEACH	RESOURCES
___ **What You Will Learn . . .** (SE) Preview Main Ideas, Big Idea, and Key Terms.	___ Daily Bellringer Transparency: Section 3
___ **Key Terms and Places** (TE) Preteach or review the key terms from this section with students.	___ RF: Vocabulary Builder: Section 3
___ **If YOU lived there . . .** (SE) Have students read and discuss the section introduction and question.	___ Differentiated Instruction Modified Worksheets and Tests CD-ROM

DIRECT TEACH	RESOURCES
___ **Teach the Big Idea Activity** (TE) Students discuss the Main Idea questions and then create a poster that includes graphs and charts.	___ Interactive Reader and Study Guide: Section 3
___ **Main Idea Questions** (TE) Discussion questions	___ RF: Map Zone Transparency: Mesopotamia and Sumer
___ **Map: Mesopotamia and Sumer** (SE)	___ RF: Geography and History: The Iraq-Iran War
___ **Differentiating Instruction: Advanced/Gifted and Talented** (TE) Hanging Gardens of Babylon	___ RF: Critical Thinking: The Country That Isn't
___ **Differentiating Instruction: English-Language Learners** (TE) Reading Activity	___ RF: Primary Source: Law of Administration for the State of Iraq for the Transitional Period, Preamble
___ **Cross-Discipline Activity: English/Language Arts** (TE) Writing About the Visuals	

REVIEW & ASSESS	RESOURCES
___ **Close** (TE) Ask students to explain the roles of war and farming in Iraq's history	___ Online Quiz Section 3 (keyword: SGD7 HP5)
___ **Section 3 Assessment** (SE)	___ PASS: Section 3 Quiz

Key: SE = Student Edition **TE** = Teacher's Edition **RF** = Resource File **PASS** = Progress Assessment Support System

The Arabian Peninsula, Iraq, and Iran

Lesson Plan

Section 4

Objectives *Students will learn that . . .*

1. Iran's history includes great empires and an Islamic republic.

2. In Iran today, Islamic religious leaders restrict the rights of most Iranians.

Vocabulary shah, revolution, Tehran, theocracy

PRETEACH	RESOURCES
___ **What You Will Learn . . .** (SE) Preview Main Ideas, Big Idea, and Key Terms. ___ **Key Terms and Places** (TE) Preteach or review the key terms from this section with students. ___ **If YOU lived there . . .** (SE) Have students read and discuss the section introduction and question.	___ Daily Bellringer Transparency: Section 4 ___ RF: Vocabulary Builder: Section 4 ___ Differentiated Instruction Modified Worksheets and Tests CD-ROM
DIRECT TEACH	**RESOURCES**
___ **Teach the Big Idea Activity** (TE) Students discuss the Main Idea questions and then develop a list of questions and answers. ___ **Main Idea Questions** (TE) Discussion questions ___ **Differentiating Instruction: Advanced/Gifted and Talented** (TE) Virtual Tour of Yazd ___ **Differentiating Instruction: Struggling Readers** (TE) Interviewing ___ **Differentiating Instruction: English-Language Learners** (TE) Chart on Iran Today ___ **Differentiating Instruction: Advanced/Gifted and Talented** (TE) Theocracy and Democracy ___ **Biography: Shirin Ebadi** (SE) ___ **Social Studies Skills: Analyzing Tables and Statistics** (SE)	___ Interactive Reader and Study Guide: Section 4 ___ RF: Biography: Masumeh Ebtekar
REVIEW & ASSESS	**RESOURCES**
___ **Close** (TE) Discuss with students the role that different rulers have had in shaping life in Iran. ___ **Section 4 Assessment** (SE)	___ Quick Facts Transparency: The Arabian Peninsula, Iraq, and Iran Visual Summary ___ Online Quiz Section 4 (keyword: SGD7 HP5) ___ PASS: Section 4 Quiz

Key: SE = Student Edition **TE** = Teacher's Edition **RF** = Resource File **PASS** = Progress Assessment Support System

Central Asia

Lesson Plan

Section 1

Objectives *Students will learn that . . .*

1. Key physical features of landlocked Central Asia include rugged mountains.

2. Central Asia has a harsh, dry climate that makes it difficult for vegetation to grow.

3. Key natural resources in Central Asia include water, oil and gas, and minerals.

Vocabulary landlocked, Pamirs, Fergana Valley, Kara-Kum, Kyzyl Kum, Aral Sea

PRETEACH	RESOURCES
___ **What You Will Learn . . .** (SE) Preview Main Ideas, Big Idea, and Key Terms. ___ **Key Terms and Places** (TE) Preteach or review the key terms from this section with students. ___ **If YOU lived there . . .** (SE) Have students read and discuss the section introduction and question.	___ Daily Bellringer Transparency: Section 1 ___ RF: Vocabulary Builder, Section 1 ___ Differentiated Instruction Modified Worksheets and Tests CD-ROM
DIRECT TEACH	**RESOURCES**
___ **Teach the Big Idea Activity** (TE) Students discuss the Main Idea questions and then create an explorer's log, noting the physical features, climate, and natural resources of Central Asia. ___ **Main Idea Questions** (TE) Discussion questions ___ **Map: Central Asia: Physical** (SE) ___ **Differentiating Instruction: English-Language Learners** (TE) Map Legends ___ **Differentiating Instruction: Advanced/Gifted and Talented** (TE) Hindu Kush Moutain Range ___ **Map: Central Asia: Land Use and Resources** (SE) ___ **Critical Thinking: Interpreting Charts** (TE) Natural Resources Chart	___ Interactive Reader and Study Guide: Section 1 ___ Map Zone Transparency: Central Asia: Physical ___ Map Zone Transparency: Central Asia: Land Use and Resources ___ RF: Geography for Life: Environmental Crises in the Aral Sea Basin ___ RF: Critical Thinking: The Death of a Sea
REVIEW & ASSESS	**RESOURCES**
___ **Close** (TE) Ask students to summarize how the physical geography of Central Asia affects people's lives there. ___ **Section 1 Assessment** (SE)	___ Online Quiz Section 1 (keyword: SGD7 HP6) ___ PASS: Section 1 Quiz

Key: SE = Student Edition **TE** = Teacher's Edition **RF** = Resource File **PASS** = Progress Assessment Support System

 Teacher Management System

Central Asia

Lesson Plan

Section 2

Objectives *Students will learn that . . .*

1. Throughout history, many different groups have conquered Central Asia.

2. Many different ethnic groups and their traditions influence culture in Central Asia.

Vocabulary Samarqand, nomads, yurt

PRETEACH	RESOURCES
___ **What You Will Learn . . .** (SE) Preview Main Ideas, Big Idea, and Key Terms. ___ **Academic Vocabulary** (TE) Review with students the high-use academic term in this section. ___ **If YOU lived there . . .** (SE) Have students read and discuss the section introduction and question.	___ Daily Bellringer Transparency: Section 2 ___ RF: Vocabulary Builder, Section 2 ___ Differentiated Instruction Modified Worksheets and Tests CD-ROM
DIRECT TEACH	**RESOURCES**
___ **Teach the Big Idea Activity** (TE) Students discuss the Main Idea questions and then create pairs of cards with information about the Silk Road. ___ **Main Idea Questions** (TE) Discussion questions ___ **Differentiating Instruction: Struggling Readers** (TE) Major Influences on Central Asia Graphic Organizer ___ **Close-up: Inside a Yurt** (SE) ___ **Critical Thinking: Comparing and Contrasting** (TE) What's in a Home? ___ **Map: Languages of Central Asia** (SE) ___ **Collaborative Learning** (TE) Creating a Map-Based Quiz	___ Interactive Reader and Study Guide: Section 2 ___ RF: Literature: "Manas" ___ Map Zone Transparency: Languages of Central Asia
REVIEW & ASSESS	**RESOURCES**
___ **Close** (TE) Ask students to explain how the different ethnic groups in Central Asia influence its culture. ___ **Section 2 Assessment** (SE)	___ Online Quiz Section 2 (keyword: SGD7 HP6) ___ PASS: Section 2 Quiz

Key: SE = Student Edition **TE** = Teacher's Edition **RF** = Resource File **PASS** = Progress Assessment Support System

Teacher Management System

Central Asia

Lesson Plan

Section 3

Objectives *Students will learn that . . .*

1. The countries of Central Asia are working to develop their economies and to improve political stability in the region.

2. The countries of Central Asia face issues and challenges related to the environment, the economy, and politics.

Vocabulary Taliban, Kabul, dryland farming, arable

PRETEACH	RESOURCES
___ **What You Will Learn . . .** (SE) Preview Main Ideas, Big Idea, and Key Terms.	___ Daily Bellringer Transparency: Section 3
___ **Key Terms and Places** (TE) Preteach or review the key terms from this section with students.	___ RF: Vocabulary Builder, Section 3
___ **If YOU lived there . . .** (SE) Have students read and discuss the section introduction and question.	___ Differentiated Instruction Modified Worksheets and Tests CD-ROM

DIRECT TEACH	RESOURCES
___ **Teach the Big Idea Activity** (TE) Students discuss the Main Idea questions and then create a flip poster of Central Asia today.	___ Interactive Reader and Study Guide: Section 3
___ **Main Idea Questions** (TE) Discussion questions	___ RF: Biography: Meena Keshwar Kamal
___ **Differentiating Instruction: Special Needs Learners** (TE) Central Asia Today Jigsaw Puzzle	___ RF: Geography and History: Afghan Refugees
___ **Focus On Culture: Turkmen Carpets** (SE)	___ RF: Primary Source: National Anthem of the Republic of Kazakhstan
___ **Collaborative Learning** (TE) Creating a TV Ad	
___ **World Almanac: Facts about Countries: Standard of Living in Central Asia** (SE)	___ RF: Biography: Al-Biruni
___ **Cross-Discipline Activity: Arts and the Humanities** (TE) Flags of Central Asia	
___ **Social Studies Skills: Using Scale** (SE)	

REVIEW & ASSESS	RESOURCES
___ **Close** (TE) Ask students for examples of how challenges in Central Asia affect specific countries there.	___ Quick Facts Transparency Central Asia Visual Summary
___ **Section 3 Assessment** (SE)	___ Online Quiz Section 3 (keyword: SGD7 HP6)
	___ PASS: Section 3 Quiz

Key: SE = Student Edition **TE** = Teacher's Edition **RF** = Resource File **PASS** = Progress Assessment Support System

                                     Teacher Management System

Southwest and Central Asia

Lesson Plans for Differentiated Instruction

The Lesson Plans for Differentiated Instruction provide teachers with strategies to help students meet benchmarks and standards.

The lesson plans provide strategies and suggestions for teaching students with special needs. Differentiated strategies are included for: (1) English-Language Instruction, (2) Special Education Instruction, (3) Advanced Learners/Gifted and Talented Instruction.

Each lesson plan is organized in two sections: (1) a main section, in which strategies appear, and (2) a side column, in which tips and side notes appear.

Main Section—Strategies: For each section in a chapter, at least one strategy is provided for at least two of the three differentiated instruction categories. Each section starts with the section number and title. The next line is the category of instruction being supported. On the next line, the title and recommended time for the strategy is provided. The title indicates how the activity might be used (such as "Prereading") or the focus of the differentiation. Following that is a brief description of the activity. If additional materials are needed for the activity, they are indicated as well. An example of a strategy format follows:

SUPPORTING SPECIAL EDUCATION INSTRUCTION
Postreading (40 minutes)

Making a Poster Have students work in small groups to create a poster about one of the topics in this section. Students' posters should include information and/or illustrations, charts, maps or graphs about Turkish history, the Ottoman Empire, or modern Turkey and its people and culture.

Side Column—Side Notes: For each section in a chapter, one or two side notes that link to a specific strategy or that support differentiated instruction for that section in general are included. These side notes are designed to give teacher tips for using the strategies or for expanding the strategies beyond what is included in the main part of the lesson plan. The side note categories are:

- Culturally Responsive Teaching
- Discipline Connection
- Grammar Tip
- Graphic Organizer
- Language Tip
- Pronunciation Tip
- Teacher Tip
- Technology Tip

Section 1: Geography of the Fertile Crescent

SUPPORTING ENGLISH-LANGUAGE INSTRUCTION
Vocabulary Analysis (15 minutes)
Identifying Additive Phrases Brainstorm with students any phrases or words that indicate additional information in the text. For example:

- in addition to

- moreover

- also

List the words on the board. Have students create a flow chart with information from the text.

SUPPORTING SPECIAL EDUCATION INSTRUCTION
Postreading (25 minutes)
Culture Areas Chart Have students create a chart of the geographical areas listed in the text by folding a piece of paper into three columns. Then have students use a ruler to draw lines between the columns. They should label the first column "area," the second column "characteristics," and the third column "civilization." Under each heading, have students list the name of each area (e.g., Mesopotamia), the characteristics (e.g., surrounded by the Tigris and Euphrates Rivers, etc.) and some information about the civilization (e.g., "hunter-gatherer groups," etc.).

SUPPORTING ADVANCED/GIFTED AND TALENTED INSTRUCTION
Interpreting an Artifact (10 minutes)
Making Inferences Look at the artifacts pictured in the chapter opener and within this section, or conduct additional research on Mesopotamian artifacts. Choose one object and think about the artisan who created it. Take on the role of the artisan and talk about the object with a partner. "What is the purpose or use of your object? To whom are you going to give your object? What were you thinking as you were working on the object? What will be your next project?" Encourage students to share their thoughts with a partner.

Resources

- Spanish Summaries Audio CD Program

- Differentiated Instruction Modified Worksheets and Tests CD-ROM:
  - Vocabulary Flash Cards
  - Vocabulary Builder
  - Section Quiz
  - Chapter Review
  - Chapter Test

Teacher Tip
Make a floor map for your classroom. Hang a shower curtain liner on the wall and project an outline map of Mesopotamia using an overhead projector. Trace the outline onto the liner with a permanent marker. Label the map with the regions and culture areas. Have students "visit" each area by walking on the map.

Technology Tip
Have students search the Internet for images of the objects from the Fertile Crescent to go along with their artisan "pair share."
Students can visit an informative Web site at:
http://oi.uchicago.edu/OI/MUS/GALLERY/EAST/New_East_Gallery.html

Teacher Management System

Section 2: The Rise of Sumer

SUPPORTING ENGLISH-LANGUAGE INSTRUCTION
Vocabulary Analysis (15 minutes)
Understanding Prefixes The text uses the word "polytheism" in its discussion of religion in Sumer. Explain to students that the prefix "poly-" means "many." List other words with the prefix on the board. For example:

polygon   polyglot   polymorphous

Ask students to list any other words they know with this prefix.

SUPPORTING SPECIAL EDUCATION INSTRUCTION
Reinforcing Vocabulary (25 minutes)
Understanding Legal Terms Have students create a personal dictionary using vocabulary words from the section. Some words they might define are:

- rural
- urban
- city-state
- empire
- priest

Encourage students to look for multiple definitions of these terms, including information from the text and from the glossary. Have students write sentences or short responses to demonstrate that they understand each term.

Section 3: Sumerian Achievements

SUPPORTING ENGLISH-LANGUAGE INSTRUCTION
Analyzing Verb Tense (10 minutes)
Understanding the Past Tense Read the following excerpt from the text to the students: "Each pictograph represented an object, such as a tree or an animal." Draw the following diagram for students:

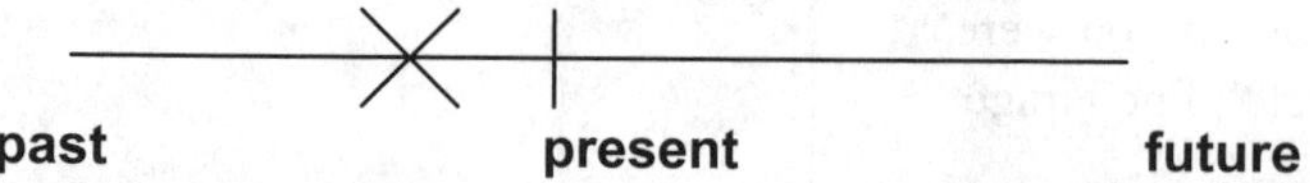

past   **present**   **future**

Explain to students that the sentence is stated in the past tense. Have students think of other examples using the same structure.

Pronunciation Tip

The word Sumer looks like the word "summer" in English, but because of the extra "m," the two words are pronounced differently. Sumer is pronounced like this:

SOO•muhr

Technology Tip

Present the information in this section to students using the computer. Use a filmstrip program or other hypercard program to arrange the main events outlined in the section. Be sure to include photos and other illustrations in your presentation.

Discipline Connection

Art: Have students view samples of pictographs on the Internet (for example, the following Web site offers photographs of Anasazi pictographs and petroglyphs: http://www.pro-visions.com/ pictogrphs.htm). After viewing examples of pictographs, have students design their own. Encourage students to share their designs with the class.

SUPPORTING SPECIAL EDUCATION INSTRUCTION
Create a Chapter Outline (40 minutes)
Sumer – the Novel Sumerian achievements are impressive. Have students brainstorm a list of the most important Sumerian achievements as described in the text. In small groups, have students design a table of contents outlining what each chapter of a book on Sumerian achievements would include. Encourage students to compare their outlines in small groups.

Section 4: Later Peoples of the Fertile Crescent

SUPPORTING ENGLISH-LANGUAGE INSTRUCTION
Vocabulary Analysis (15 minutes)
Identifying Changing Phrases Brainstorm with students any phrases or words that indicate a change of information in the text. For example: "<u>Although</u> Ur rose to glory after the death of Sargon, repeated foreign attacks drained its strength." Some words to highlight are: *nonetheless, despite, however, although.* Write the list of words on the board. Have students create a comparison chart like the one at right with the information that comes from the text.

SUPPORTING ADVANCED/GIFTED AND TALENTED INSTRUCTION
Interpreting Information (15 minutes)
Writing a Newspaper Article Have students imagine they just heard about Hammurabi's Code. Have them brainstorm a list of words that describe the Code and the types of laws covered in the Code. With this information, have students write a short newspaper article announcing the new Code and describing some of the laws. They should begin their article with a headline (e.g., "Hammurabi shocks Mesopotamia with his new Code"). The first line should summarize the event (e.g., "Hammurabi, respected leader and ruler, presented his new Code to the Babylonian people today. The laws were received with rave reviews…"). Encourage students to include as much information as possible in their articles.

Teacher Tip
Have students refer to the table of contents of a nonfiction book, pointing out how the list of chapters provides a brief line of information about each one.

Graphic Organizer
Comparison Chart

Fact 1	Fact 2
"Ur rose to glory after the death of Sargon."	"Repeated foreign attacks drained its strength."

Language Tip
Writing for a newspaper article is very different from the writing that goes into an essay or composition. Have students brainstorm the differences in each genre of writing.

Section 1: Origins of Judaism

SUPPORTING ENGLISH-LANGUAGE INSTRUCTION
Prereading (10 minutes)
Analyzing Text Structure: Sequential Narration Explain to students that the early history of the origins of Judaism in this section of the text is narrated in a sequential, or chronological, fashion according to the dates of each event. Have students look through the text under the subhead "Early History" and write a list of the dates that appear. They should list these dates down the left side of a sheet of paper.

Reading (40 minutes)
Analyzing Text Structure: Sequential Narration As students read, have them write a brief sentence or two to describe what happened on each of the dates they listed in the previous activity. When students complete their lists, have them create a time line with the information.

SUPPORTING SPECIAL EDUCATION INSTRUCTION
Postreading (25 minutes)
Religious Texts Chart Have students create a chart of the Jewish texts listed in this section by folding a piece of paper into three columns. Then have students use a ruler to draw lines between the columns. They should label the first column "The Torah," the second column, "The Hebrew Bible," and the third column "The Commentaries." Under each heading, have students list some important characteristics of each text.

SUPPORTING ADVANCED/GIFTED AND TALENTED INSTRUCTION
Synthesizing Information (40 minutes)
Creating an Advertising Poster In this section students learn about Moses, a Hebrew leader who tried to convince the pharaoh to free the Hebrews from slavery. Have students imagine that they are Moses and ask them to design an advertising poster with the same goal in mind. Encourage them to include information from the text, as well as an interesting heading and bulleted phrases to catch the reader's attention.

Resources
- Spanish Summaries Audio CD Program
- Differentiated Instruction Modified Worksheets and Tests CD-ROM:
  –Vocabulary Flash Cards
  –Vocabulary Builder
  –Section Quiz
  –Chapter Review
  –Chapter Test

Culturally Responsive Teaching
Have students read the text under the subheading "Jewish Beliefs" in this section. Ask them to think about similar beliefs or values in other religious faiths or cultural traditions they may be familiar with.

Technology Tip
Have students do further research on the origins of Judaism at the following Web site: http://www.pbs.org/wnet/heritage/episode1/

The site contains an interactive atlas, historical documents, and video clips from the PBS series, *Heritage: Civilization and the Jews*.

Section 2: Origins of Christianity

SUPPORTING ENGLISH-LANGUAGE INSTRUCTION
Vocabulary Analysis (15 minutes)
Identifying Time Words Write a list of time phrases for students on the board including:

• soon

• eventually

• gradually

• after

• first

• second

Have students find these words in the section. Ask them to create a time line of the events described based on the use of these words.

SUPPORTING SPECIAL EDUCATION INSTRUCTION
Spreading Christianity (30 minutes)
Role Play Explain to students that most descriptions in this section are written in third person (e.g., *He* traveled . . .). Then explain that when a person speaks directly about his/her experiences, we use first person (e.g., *I* traveled . . .). In small groups, have students write a short skit about spreading the teachings of Christianity. Students should write the script in the first person, taking on the roles of the Apostles. Skits should include information about the Gospels, Jesus's message, and where and how Christianity spread. Encourage students to be creative.

SUPPORTING ADVANCED/GIFTED AND TALENTED INSTRUCTION
Synthesizing Information (30 minutes)
Analyzing Images Have students look at the depictions of the origins of Christianity included in this section, such as the scene from Jesus's childhood and the Last Supper. Ask students to choose an image and think about the artisan who created it. Students should take on the role of the artisan and discuss the image with a partner. Possible discussion topics include: "What were you thinking when you were working on the image? To whom are you going to give your image? What will your next project be?"

Teacher Tip
Begin the lesson by "previewing" some of the key vocabulary from the section. Find illustrations or photos to represent the vocabulary. Place these items in a bag or box with a lid. Have students pull the photos or illustrations from the bag one by one and discuss them. Then have students provide a short definition or explanation.

Technology Tip
Have students research different theories about the origins of the four Gospels at the following Web site:
http://www.pbs.org/wgbh/pages/frontline/shows/religion/story/mmindex.html

Section 3: The Byzantine Empire

SUPPORTING ENGLISH-LANGUAGE INSTRUCTION
Vocabulary Analysis (15 minutes)
Understanding Suffixes Read the following excerpt from the text to students: "Byzantine emperors had more power than western emperors did." Explain to students that an adjective can be formed by adding the suffix "-ine" to a noun:

noun + -ine = adjective

Byzanthium + -ine = Byzantine

List other words with the suffix on the board. Ask students to list any other words they know with this suffix.

SUPPORTING SPECIAL EDUCATION INSTRUCTION
Interpreting Information (25 minutes)
Writing a Letter Have students imagine that they are merchants who have traveled to Constantinople from Europe, Asia, or Africa to trade. Have them brainstorm a list of words that describe the city, how they feel, and what they see and hear. With this information, have students write a short letter to a friend about the experience. Have students exchange their letters with other members of the class.

SUPPORTING ADVANCED/GIFTED AND TALENTED INSTRUCTION
Synthesizing Information (30 minutes)
Justinian—the Movie The life of Justinian has all the elements of a movie—action, drama, and battles. Have students choose a detail about Justinian's life and brainstorm what a movie about his experiences as emperor might be like. Have them design a storyboard with several scenes that might appear in the film. Students should include characters, a setting, and sample dialogue for each scene.

Teacher Tip
The word "Constantinople" is pronounced like this:

kahn • stan • ti • NO • puhl

Grammar Tip
Remind students that verbs carry out the action of a sentence. Transitive verbs are verbs that require a direct object to receive their action.

Discipline Connection
Art: Have students view Byzantine art at the following Web site:
http://www.rom.on.ca/galleries/byzantine/byzindex.html

Section 1: Origins of Islam

SUPPORTING ENGLISH-LANGUAGE INSTRUCTION
Vocabulary Analysis (15 minutes)
Identifying Changing Phrases Brainstorm with students any phrases or words that indicate a change of information in the text. For example:

- nonetheless
- despite
- however

List the words on the board. Have students create a comparison chart with information from the text.

SUPPORTING SPECIAL EDUCATION INSTRUCTION
Reinforcing Vocabulary (25 minutes)
Understanding Religious Terms Have students create a personal dictionary using vocabulary words from the section. Some words they might define are: Muhammad, Islam, Muslim, Qur´an, pilgrimage, and mosque. Encourage students to look for multiple definitions of these terms, including information from the text and from the glossary. Have students write sentences or short responses to the terms to demonstrate that they understand each one.

SUPPORTING SPECIAL EDUCATION INSTRUCTION
Reviewing Information (20 minutes)
Self-Questioning Techniques Have students use self-questioning techniques to review the material addressed in this section. As they read, encourage students to write down questions about the material that come up in the reading. Students should then meet in small groups to share their questions and look for answers in their textbooks.

SUPPORTING ADVANCED/GIFTED AND TALENTED INSTRUCTION
Synthesizing Information (30 minutes)
Writing a Compare/Contrast Essay So far, students have read about nomads and townspeople. Have them write an essay about these two groups in which they discuss the similarities and differences between them. Encourage students to state whether they would want to live in town or as a nomad. Ask students to relate their writing to anything they may have seen, heard, or read in the past that gives them an idea about the topic. Remind students that they may illustrate their story if they wish.

Resources
- Spanish Summaries Audio CD Program
- Differentiated Instruction Modified Worksheets and Tests CD-ROM:
  - Vocabulary Flash Cards
  - Vocabulary Builder
  - Section Quiz
  - Chapter Review
  - Chapter Test

Language Tip
Explain to students that, since most of the terms regarding Islam come from Arabic, a language written in a different script, the words can be translated many different ways when written in English. Here are some alternative spellings that you might see in other books:

- Muhammad = Mohammed
- Muslim = Moslem
- Qur´an = Koran

Although translations of the Qur´an exist in many world languages, only Arabic versions of the holy book are considered sacred by Muslims because the Qur´an was originally dictated to Muhammad in Arabic.

Section 2: Islamic Beliefs and Practices

SUPPORTING ENGLISH-LANGUAGE INSTRUCTION
Vocabulary Analysis (15 minutes)
Identifying Ordinal Numbers Write a list of ordinal numbers in both words and numerals for students on the board including: first = 1st, second = 2nd, third = 3rd, fourth = 4th, fifth = 5th. Have students write the numerals down the left-hand side of a sheet of paper. Then have them list the pillars of Islam next to each number.

SUPPORTING SPECIAL EDUCATION INSTRUCTION
Comparing Information (30 minutes)
A Behavior Chart Have students create a chart with information about the Islamic rules for behavior outlined in this section. Have students brainstorm the rules and then write them down the left-hand side of a sheet of paper. Ask students to check each rule that is also considered important in their own culture or faith.

SUPPORTING ADVANCED/GIFTED AND TALENTED INSTRUCTION
Synthesizing Information (30 minutes)
A Legal Debate Read the following sentences to the students: "Shariah deals with punishments for people who break society's rules. For example, Shariah says Muslims must not steal and that getting a hand cut off is the punishment for stealing." Remind students that, throughout history, this type of punishment is common (e.g., Hammurabi's code of "an eye for an eye"). Ask students to debate the legal issues surrounding this idea. Have students divide into two groups and formulate their opinions and ideas about the topic. Then ask students to debate both sides of the issue.

Teacher Tip
Provide students with a copy of your lecture or presentation notes for their review and for use as a study guide.

Technology Tip
Present the information in this section to students using the computer. Use a filmstrip or other hypercard program to arrange the main events outlined in the section. Be sure to include photos and other illustrations in your presentation.

Teacher Tip
Before students read, give them a list of questions that they should be able to answer once they have read the section. Check for comprehension after they read by going over the questions with the class.

Language Tip
The word "pilgrimage" is formed with the root "pilgrim," a word that means "foreigner," or "from abroad."

Teacher Management System

Section 3: Muslim Empires

SUPPORTING ENGLISH-LANGUAGE INSTRUCTION
Vocabulary Analysis (15 minutes)
Understanding Prefixes In this section of the text, students read about *conquest* and later will read about Islamic *contributions*. Explain to students that the prefix *con-* means *with* or *together*. List other words with the prefix on the board. For example:

conversation   contact

Ask students to list any other words they know with this prefix.

SUPPORTING ADVANCED/GIFTED AND TALENTED INSTRUCTION
Synthesizing Information (30 minutes)
Varieties of Islam Have students research the three main branches of Islam, the Sunni, the Sufi, and the Shia Muslims. Ask students to create a poster or a pamphlet describing the similarities and differences between the three branches, noting any smaller subgroups within each branch, and indicating in what countries each group can be found. Students should explain how the beliefs of each group affect how it governs and influence the group's relationships with other Muslims and with non-Muslims.

Section 4: Cultural Achievements

SUPPORTING SPECIAL EDUCATION INSTRUCTION
Synthesizing Information (30 minutes)
Role-Play Being a Patron Tell students that they are wealthy Muslims who want to commission a new building. Have students describe or draw the building they would like to have built. Students should explain the use of the building and give some idea of the floor plan of the building. Encourage students to share their building ideas with the class.

SUPPORTING ADVANCED/GIFTED AND TALENTED INSTRUCTION
Interpreting Information (30 minutes)
Analyzing Visuals Have students read about Muslim inventions in this section. Ask them to choose one invention and think about the inventor who created it. Have the student take on the role of the inventor and talk about the object with a partner. "What is the purpose of the invention? Who will benefit most from the invention? What gave you the idea for this invention? What will be your next project?" Have students present their thoughts to the class.

Discipline Connection
Art: Have students investigate examples of Mughal art on the Internet. One useful Web site can be accessed at the following address:
http://www.infoplease.com/ce6/ent/A0834336.html

Teacher Tip
Make a floor map for your classroom. Hang a shower curtain liner on the wall and project an outline map of the Muslim world using an overhead projector. Trace the outline onto the liner with a permanent marker. Label the map with the countries and important cities of trade. Have students "trade goods" with one another by walking on the map from country to country.

Discipline Connection
Science: Have students investigate Islamic scientific and mathematical contributions to the world. An excellent source for information can be found on the Internet at the following Web address:
http://www.sfusd.k12.ca.us/schwww/sch618/ScienceMath/Science_and_Math.html

Teacher Tip
Provide students with the reading in smaller chunks. You can do this by assigning students to read each sub-section separately.

Teacher Management System

Section 1: Physical Geography

SUPPORTING ENGLISH-LANGUAGE INSTRUCTION
Vocabulary Analysis (20 minutes)
Understanding Key Terms Have students list the key terms on the left side of a sheet of paper. With a partner, have students describe the terms and/or draw a picture on the right side of the paper. Then have students use library or Internet resources to find out more about each term. Encourage students to share their findings with the class.

SUPPORTING ADVANCED/GIFTED AND TALENTED INSTRUCTION
Synthesizing Information (40 minutes)
Creating a Diorama Have students create a diorama of a desert scene in either the Syrian or Negev deserts. Students' dioramas should include a representative sampling of plant, animal, and human life in the deserts.

Section 2: Turkey

SUPPORTING SPECIAL EDUCATION INSTRUCTION
Postreading (40 minutes)
Making a Poster Have students work in small groups to create a poster about one of the topics in this section. Students' posters should include information and/or illustrations, charts, maps, or graphs about Turkish history, the Ottoman Empire, or modern Turkey and its people and culture.

SUPPORTING ADVANCED/GIFTED AND TALENTED INSTRUCTION
Synthesizing Information (45 minutes)
Making Connections to Modern Times After reading about Turkey's history of invasions by outsiders, have students think about what a similar situation might be like in modern times. Encourage students to think about how they would feel if invaders continually attacked their country. Have students write a skit in which they rebel against modern-day invaders. Ask students to perform their skit in front of the class.

Resources
- Spanish Summaries Audio CD Program
- Differentiated Instruction Modified Worksheets and Tests CD-ROM:
  - Vocabulary Flash Cards
  - Vocabulary Builder
  - Section Quiz
  - Chapter Review
  - Chapter Test

Technology Tip
Help students with their dioramas by directing them to information about deserts at the following Web site:
http://pubs.usgs.gov/gip/deserts/contents/

Discipline Connection
Art: Have students research Turkish art and culture at the following Web address:
http://www.kultur.gov.tr/portal/default_en.asp

Language Tip
Writing a skit or for a poster is very different from the writing that goes into an essay. Have students brainstorm the difference in each genre of writing.

Teacher Management System

Section 3: Israel

SUPPORTING ENGLISH-LANGUAGE INSTRUCTION
Analyzing Verb Tense (10 minutes)
Understanding the Present The present tense is used in this section to discuss Israel today (for example: "Israel has a prime minister and a parliament—the Knesset"). The present describes a fact that was true in the past, is true now, and will be true in the future. Have students write sentences that describe Israel in the past, in the present, and in the future.

SUPPORTING SPECIAL EDUCATION INSTRUCTION
Reading (40 minutes)
Self-Questioning Techniques Have students use self-questioning techniques to review the material in this section. As students read, encourage them to write their questions on a separate sheet of paper. Students should then meet in small groups to share their questions and look for answers in their textbooks.

Section 4: Syria, Lebanon, and Jordan

SUPPORTING ENGLISH-LANGUAGE INSTRUCTION
Postreading (25 minutes)
Border Countries Chart Have students create a chart of the different countries listed in the text that border Israel by folding a piece of paper into three columns. Then have students use a ruler to draw lines between the columns. They should label the first column "country," the second column "government," and the third column "people." Under each heading, have students list the name of each border country (e.g., Syria), some information about its government (e.g., was led by Hafiz al-Assad until 2000), and some information about its people (e.g., "about 90 percent Arab," etc.).

SUPPORTING ADVANCED/GIFTED AND TALENTED INSTRUCTION
Synthesizing Information (30 minutes)
Creating a TV Newscast Have students write a TV newscast about Lebanon's civil war as described in the reading. In their scripts, they should include reasons for the civil war, the different ethnic groups involved, and how long the civil war lasted. Students should also create or use the Internet to find visuals (e.g., a chart or a photo) for use in their newscasts.

Discipline Connection
English/Language Arts: Have students research the Dead Sea Scrolls. A useful Web site can be found at the following address: http://www.ibiblio.org/expo/ deadsea.scrolls.exhibit/intro.html

Pronunciation Tip
The word "Knesset" is pronounced like this:

knes • et

Culturally Responsive Teaching
Have students share details about their own ethnic backgrounds. Encourage students to describe a particular custom or bring to class a picture or sample of traditional clothing, food, music, or art.

Teacher Tip
Have students watch a newscast on TV or listen to a newscast on the radio to help them become familiar with the writing style.

Teacher Management System

Section 1: Physical Geography

SUPPORTING ENGLISH-LANGUAGE INSTRUCTION
Understanding Vocabulary (15 minutes)
Analyzing Compound Words The text uses the word "groundwater" to describe water in wells that rainfall does not replace. Explain to students that combining two nouns to express a single idea forms many compound words. List other compound words from the section on the board. For example:

landform   nighttime   underground

Ask students to list any other compound words they know.

SUPPORTING SPECIAL EDUCATION INSTRUCTION
Prereading (10 minutes)
Predicting the Main Idea Before students read, have them look at the headings, subheadings, and highlighted words in this section. They should also pay attention to the maps and other graphics that appear alongside the reading. Based on this information, ask them to predict the main idea of the section. Have them share their predictions with a partner. As they read, encourage students to determine whether their predictions were correct.

Reading (30 minutes)
Identifying Difficult Passages As students read the section for the first time, have them place sticky notes next to passages that are confusing or that they do not understand. On each sticky note, have students write a question or a vocabulary word with which they are having difficulty. Once they finish reading, help them to answer their questions by directing them to related Web sites, dictionaries, or by conferencing with them. Have students re-read the passage to check the answers they receive.

SUPPORTING ADVANCED/GIFTED AND TALENTED INSTRUCTION
Synthesizing Information (40 minutes)
Creating a Travelogue After reading about the Arabian Peninsula, have students imagine that they are traveling through the Rub´ al-Khali Desert to the Red Sea. Have them write a travelogue about their experiences based on the information in the section. Students should also create or use the Internet to find visuals (e.g., a map or a photo) for use in their travelogues.

Resources
- Spanish Summaries Audio CD Program
- Differentiated Instruction Modified Worksheets and Tests CD-ROM:
  - Vocabulary Flash Cards
  - Vocabulary Builder
  - Section Quiz
  - Chapter Review
  - Chapter Test

Teacher Tip
Explain to students that a travelogue can be a talk, a film, or slides that accompany a piece of writing about travel. Tell students that they can use photos or maps from the Internet, or their own drawings if they wish.

Technology Tip
Help students with their travelogues by directing them to information and photos of the Rub´ al-Khali Desert at the following Web site:
http://seabed.nationalgeographic.com/studentatlas/clickup/deserts.html

Teacher Management System

Section 2: The Arabian Peninsula

SUPPORTING ENGLISH-LANGUAGE INSTRUCTION
Prereading (10 minutes)
Analyzing Text Structure: Descriptive Narration Explain to students that the information about the countries of the Arabian Peninsula in this section features descriptive narration. Have students look through the text and write a list of the main points presented in this section about Saudi Arabia and the other countries of the Arabian Peninsula. Students should list these points down the left-hand side of a sheet of paper.

Reading (10 minutes)
Analyzing Text Structure: Descriptive Narration As students read, have them write a brief sentence or two to describe each point they listed in the previous activity. When students have completed their lists, have them create an outline with the information.

SUPPORTING SPECIAL EDUCATION INSTRUCTION
Interpreting Information (25 minutes)
Writing a Letter Have students imagine that they are visiting Saudi Arabia or one of the other countries in the Arabian Peninsula. Have them brainstorm a list of words to describe how they feel and what they see and hear. With this information, have students write a short letter to a friend about their trip. Have students exchange letters with each other in class.

SUPPORTING ADVANCED/GIFTED AND TALENTED INSTRUCTION
Role Play (40 minutes)
Writing an Interview In this section, students read about the king of Saudi Arabia. Have them conduct research to learn more about his life. Then ask students to write an interview with him. Remind students that they are taking on the role of the interviewer and the role of the king.

Culturally Responsive Teaching
Have students read about the Qur´an in this section of the text. Ask them to think about similar holy books or teachings in their own faiths or cultural traditions. Then have students research different cultural traditions at the following Web site:
http://www.mnsu.edu/emuseum/cultural/religion/

Pronunciation Tip
The word "Sunni" is pronounced like this:

SOO • nee

Language Tip
Writing lines for a role play is very different from writing an essay or a composition. Have students brainstorm the differences in each genre of writing.

Section 3: Iraq

SUPPORTING ENGLISH-LANGUAGE INSTRUCTION
Paraphrasing (30 minutes)
Iraq Today After reading the section about Iraq today, have students brainstorm a list of words that describe life in present-day Iraq. Students should then use this information to write a summary paragraph about Iraq today. Encourage students to share their paragraphs with a partner.

SUPPORTING ADVANCED/GIFTED AND TALENTED INSTRUCTION
Interpreting Information (30 minutes)
Creating a Political Cartoon After reading about Saddam Hussein, have students design a political cartoon that illustrates his rule. They should include some form of illustration (encourage students to use computer graphics if they prefer not to draw) along with a speech bubble and/or caption. Remind students that their cartoon does not necessarily have to be funny. It should, however, depict Saddam Hussein's rule in some novel way. Have students share their cartoons with the class.

Section 4: Iran

SUPPORTING ENGLISH-LANGUAGE INSTRUCTION
Vocabulary Analysis (15 minutes)
Understanding Suffixes Read the following excerpt from the text to students: "The current government of Iran is a theocracy—a government ruled by religious leaders." Explain to students that the suffix -*cracy* means a form of government. Ask students to list any other words they know with this suffix.

SUPPORTING SPECIAL EDUCATION INSTRUCTION
Making Lists (15 minutes)
Society in Iran After students read the passages about Iran today, have them make a bulleted list of some of the main characteristics of Iranian society. After creating the list, have students put a check next to those elements that are also part of our society.

Grammar Tip
Remind students that adding the letter "s" to the end of a noun creates the plural form of most nouns. Words that end in -*ch*, *x*, *s*, or *s*-like sounds need "es" to become plural (e.g., branches).

Teacher Tip
Use current newspaper or magazine articles about Iraq to update students on life in Iraq today.

Discipline Connection
Art: Have students research Persian Empire art and architecture at the following Web site:
http://www.bartleby.com/65/pe/Persiana.html

Technology Tip
Have students research interesting facts about Shirin Ebadi at the following Web site:
http://nobelprize.org/peace/laureates/2003/ebadi-bio.html

Section 1: Physical Geography

SUPPORTING ENGLISH-LANGUAGE INSTRUCTION
Vocabulary Analysis (15 minutes)

Understanding Prefixes The text uses the word "landlocked" to describe the countries in Central Asia. Explain to students that combining two nouns to express a single idea forms many compound words. List other compound words on the board. For example:

freshwater   *rainfall*   *pipelines*

Ask students to list any other compound words they know.

SUPPORTING SPECIAL EDUCATION INSTRUCTION
Reading (20 minutes)

Taking Notes As students read, have them take notes about the important topics and key vocabulary words by making a bulleted list of phrases. Remind students that they do not have to write full sentences but rather key words and ideas that they think are important to remember. Remind students that the words and phrases in bold in the textbook are important to know, but that they should also include any other ideas they think might be important.

SUPPORTING ADVANCED/GIFTED AND TALENTED INSTRUCTION
Synthesizing Information (40 minutes)

Creating a Diorama Have students create a diorama of either a mountain or a desert scene in Central Asia. Students' dioramas should reflect the physical geography of the region as described in the reading, and include representations of some of the key natural resources found in the region.

Resources

- Spanish Summaries Audio CD Program

- Differentiated Instruction Modified Worksheets and Tests CD-ROM:
  - Vocabulary Flash Cards
  - Vocabulary Builder
  - Section Quiz
  - Chapter Review
  - Chapter Test

Pronunciation Tip

The country Kyrgyzstan is pronounced like this:

kir • gi • stan

Technology Tip

Help students with their dioramas by directing them to information about the Pamirs and the Kara-Kum Desert at the following Web sites:

http://www.pamirs.org/

http://enrin.grida.no/htmls/turkmen/soe2/index.htm

Section 2: History and Culture

SUPPORTING ENGLISH-LANGUAGE INSTRUCTION
Vocabulary Analysis (15 minutes)
Understanding Suffixes Read the following excerpt from the text to students: "A yurt is a movable round house made of wool felt mats hung over a wood frame." Explain to students that an adjective can be formed by adding the suffix "-able" to a noun:

$$noun + \text{-able} = adjective$$
$$move + \text{-able} = movable$$

Ask students to list any other words they know with this suffix.

SUPPORTING SPECIAL EDUCATION INSTRUCTION
Paraphrasing (30 minutes)
People in Central Asia After reading the section about culture in Central Asia, have students brainstorm a list of words and phrases that describe the people of Central Asia. Students should then use this information to write a summary paragraph about the different peoples in the region, including their languages and religions.

Section 3: Central Asia Today

SUPPORTING ENGLISH-LANGUAGE INSTRUCTION
Reading (40 minutes)
Analyzing Text Structure: Cause and Effect Narration Have students create a T-chart with the word "cause" written at the top of one column and the word "effect" at the top of the other column. As students read, have them copy down any phrases that indicate effect. When they have finished reading, have them fill in the causes for each effect.

Cause	Effect
Kazakhstan was the first part of Central Asia to be conquered by Russia.	As a result, Russian influence remains strong in the country today.

SUPPORTING ADVANCED/GIFTED AND TALENTED INSTRUCTION
Interpreting Information (30 minutes)
Writing a Compare/Contrast Essay So far students have read about several different countries in Central Asia. Have them write an essay about two of these countries in which they discuss the similarities and differences between them. As a conclusion, have students summarize their thoughts about the importance of these two countries to the culture of Central Asia.

Culturally Responsive Teaching
Have students share details about their own ethnic backgrounds. Ask students if their native language is the primary language for their country, and whether people in their country speak more than one language. Students can research the languages in their native country at the following Web site:

http://www.infoplease.com/ipa/A0855611.html

Grammar Tip
Remind students that the words *a, an,* and *the* are called articles. Articles are a type of adjective that answer the question *Which one? The* is called the *definite article* because it usually comes before a specific noun (e.g., *the* Taliban). *A* and *an* are called *indefinite articles* because they usually refer to less specific nouns (e.g., *a* group of reformers, *an* elected president).

Discipline Connection
Art: Have students view photos of Turkmen carpets at the following Web site:

http://www.turkishculture.org/tapestry/turkistan2.html

Teacher Management System

Southwest and Central Asia

Interactive Reader and Study Guide

The *Interactive Reader and Study Guide* provides a chapter reading strategy and summaries of the main ideas of each section. The goal of this *Intereactive Reader and Study Guide* is to provide a written summary of each section in the textbook for students who are having problems accessing the concepts that they must master for assessment. The guide should help students understand the big ideas or main concepts of each section.

The *Interactive Reader and Study Guide* provides students with opportunities to interact with the text, which allows them to access the power and meaning of the printed word. Interaction with the text should help students develop positive attitudes about reading.

The structure of the activities allows students to warm up by looking at a chapter summary graphic organizer that summarizes the main ideas in the chapter. These organizers are based on the Big Ideas and Main Ideas listed at the start of each section.

Each graphic organizer is followed by two to four Comprehension and Critical Thinking questions. These questions are related to the material in the graphic organizer. Questions frequently direct students to complete part of the graphic organizer or to add to it. Sometimes, students are instructed to complete the graphic organizer and then answer the questions.

Each of the section summaries has features to help students access the main ideas of the section. These features include:

- **Main Ideas Box** This is the text of the section main ideas from the section opener page in the Student Edition chapter.

- **Key Terms and Places** This is the list of key terms and places from the Student Edition section opener page. Each term has a short definition that matches the definitions provided in the running text of the Student Edition. The definitions have been shortened if necessary.

- **Section Summary** This is a summary of the section text. The summary covers the section's main ideas, addresses the power standard, and includes all boldfaced key terms and places in the section. The section heads from the student text are used to organize the summary.

- **Call-Out Box Questions and Activities** These activities appear in the side column next to the summary. The questions and activities are written to encourage students to interact with the text. These questions and activities may provide additional information related to a portion of the summary, ask students to circle or underline specific parts of the summary, or ask questions about the summary. The questions and activities vary both in type and difficulty level. They help reinforce, explain, clarify, or extend the main ideas in the section. These questions and activities are designed to be keys to help students unlock the text—and thereby help students who have difficulty accessing the concepts to master the information they need for assessment.

- **Challenge Activity** This activity follows the section summary. It is a short activity that challenges students and builds off of one or more of the section's main ideas. This Challenge Activity is usually aimed at advanced and gifted and talented students.

Teacher Management System

When students have finished the Chapter Opener and Section Summary pages for a chapter, they should have a grasp of the main ideas in the section—the concepts that will be assessed on standardized tests. The *Interactive Reader and Study Guide* can be used by a teacher in a variety of ways, including

- as a way to review and reinforce the regular student textbook,

- as a way to give struggling students more time to focus on only the most important parts of the section, which will give them the chance to master the material and be successful when assessed, and

- as a way to give advanced or gifted and talented students an opportunity to master the basic material quickly so that they might move on to an extension activity.

History of the Fertile Crescent

COMPREHENSION AND CRITICAL THINKING

1. farming, cities; Students may also present logical arguments to support "social hierarchy" or "religion" as economic structures.
2. the arts, writing, trade
3. Accept answers that students can support, such as farming, religion, and invention.

SECTION 1
Call-Out Boxes

1. the Tigris and Euphrates rivers
2. wheat, barley
3. Flooding destroyed crops, killed livestock, and washed away homes. When water levels were too low, crops dried up.
4. from rivers
5. cities

Challenge Activity

Students' proposals should include the irrigation methods mentioned in the chapter: storage basins, canals, a network of ditches to bring water to fields and grazing areas.

SECTION 2
Call-Out Boxes

1. more people lived in the cities
2. about 700 miles
3. Answers will vary but should demonstrate an understanding of the role that religion can play in public life today.
4. They made offerings to them.
5. priests and nobles

Challenge Activity

Students' letters should reflect an awareness of the king's and priests' positions within the social hierarchy of Sumer.

SECTION 3
Call-Out Boxes

1. cuneiform
2. picture symbols that represent objects
3. the ox-drawn plow
4. Most people lived in one-story houses with rooms arranged around a small courtyard.
5. stringed instruments, reed pipes, drums, and tambourines

Challenge Activity

possible answers—writing: this list, classwork, homework; the wheel: travel to school, riding a bike, skating

SECTION 4
Call-Out Boxes

1. Euphrates River
2. so everyone could see the laws
3. Hittites, Kassites, Assyrians, and Chaldeans
4. Sumerian civilization
5. Mediterranean Sea

Challenge Activity

Students' time lines should include the following: 1800 BC—Babylon rises on the Euphrates, 1595 BC—Hittites of Asia Minor capture Babyon, 1600 BC–1200 BC—Kassites rule Babyon, 1200 BC—Assyrians conquer Mesopotamia, 612 BC—Chaldeans drive Assyrians from power

Judaism and Christianity

COMPREHENSION AND CRITICAL THINKING

1. Christianity
2. Judaism
3. Gospels written by Apostles

SECTION 1
Call-Out Boxes

1. Abraham
2. the scattering of Jews outside of Israel
3. there is one and only one God
4. by studying Judaism's sacred texts and honoring Jewish traditions and holy days

Challenge Activity

Students' recommendations will vary. Have them give reasons for their suggestions based on which aspect of Jewish history they find most interesting.

Teacher Management System

SECTION 2
Call-Out Boxes

1. his teachings challenged the authority of Roman leaders
2. that he rose from the dead, appeared to his disciples, gave them instructions, and then rose up to heaven
3. to love God and love other people; salvation
4. possible answer—He spread Jesus's teachings.
5. Constantine

Challenge Activity

Letters will vary but should include detailed information on the long-term effects of Paul's ministry.

SECTION 3
Call-Out Boxes

1. a legal system with simplified Roman laws
2. Theodora
3. 1453; Constantinople was conquered by invaders.
4. It had non-Roman influences; its emperors were heads of the church as well as political leaders.
5. These were pictures made with pieces of colored stone or glass. Some were made of gold, silver, and jewels.

Challenge Activity

Diagrams will vary. Students should mention non-Roman influences in the eastern empire in addition to the different kinds of power wielded by the eastern and western emperors.

History of the Islamic World

COMPREHENSION AND CRITICAL THINKING

1. Abu Bakr; he unified Arabia for the first time.
2. Muhammad's teachings and the introduction of Islam to Arabia required changes in how rich and poor interacted, outlawed slavery, and introduced other cultures as the religion spread.

3. possible answers—advances in astronomy (improving the astrolabe), mathematics (algebra), and medicine (Ibn-Sina's medical encyclopedia)
4. No, women and non-Muslims did not have the same rights as Muslim men.

SECTION 1
Call-Out Boxes

1. Africa, Europe, and Asia
2. for protection and to reduce competition for grazing lands
3. Muhammad
4. in a cave
5. possible answer—They might have wanted to keep the money for themselves and their families.

Challenge Activity

possible answers—Nomadic life: constant travel, raising animals, association with a small, tight-knit tribe or group; Sedentary life: stable home base, living in a city or town, and working at a shop or trade

SECTION 2
Call-Out Boxes

1. the Qur´an
2. possible answer—It could be interpreted to mean to defend the Muslim community against attacks from non-Muslims.
3. No, it is a collection of Muhammad's words and actions.
4. a yearly donation to charity
5. No, most Muslim countries blend Shariah with a legal system similar to that in the United States.

Challenge Activity

possible similarities—belief in one god, rules of behavior defined by god, and responsibility for the welfare of others; possible differences—specific prayer requirements, influence of Shariah law, and the requirement that believers make a pilgrimage.

Teacher Management System

SECTION 3
Call-Out Boxes
1. from India to Spain
2. Goods can travel by boat to and from coastal cities more easily than to inland cities.
3. boys from conquered towns who were enslaved and converted to Islam
4. 1453
5. Yes, conflict still exists today between the Sunnis and the Shias.

Challenge Activity
Time line/map should include several key conquests: areas in Northern Africa, Spain, and France, to Northern India prior to 1000; southern Byzantine areas in mid-1200s; Constantinople/Turkey in 1453 and eastern Mediterranean into Europe in 1500s

SECTION 4
Call-Out Boxes
1. Baghdad and Córdoba
2. Astrolabes could help sailors navigate by the stars.
3. how to detect and treat smallpox, a medical encyclopedia
4. *The Thousand and One Nights*
5. Muslims believe only Allah can create humans and animals or their images.

Challenge Activity
Answers will vary and may include any of the advances mentioned in mathematics, astronomy, medicine, and the arts. Students should explain why the choice is the most important to modern society.

The Eastern Mediterranean

COMPREHENSION AND CRITICAL THINKING
1. under Lebanon—History; under Israel—People/Culture
2. basalt, kibbutz
3. Lebanon, Israel, Jordan, Syria, Turkey

SECTION 1
Call-Out Boxes
1. the Dardanelles, the Bosporus, and the Sea of Marmara
2. the Pontic and the Taurus Mountains
3. dry, Mediterranean climate, desert climate, steppe climate, humid subtropical climate
4. sulfur, mercury, and copper

Challenge Activity
Students may say that the best location for farming would be in the coastal areas of Turkey and northern Israel because they have a Mediterranean climate.

SECTION 2
Call-Out Boxes
1. They named it after their emperor, Constantine.
2. a nomadic people from Central Asia who invaded the area of Turkey in the AD 1000s
3. Mustafa Kemal
4. He believed modernizing Turkey would make it a strong nation.
5. Kurds
6. Turkey's legislature

Challenge Activity
Turkey's economy would benefit by increased trade with European countries. It would bring more money into the country and improve the daily lives of the people living there.

SECTION 3
Call-Out Boxes
1. It is home to sacred sites of three major religions: Judaism, Islam, and Christianity.
2. the 60s BC
3. about 80 percent
4. large farms where people share everything in common
5. Gaza and the West Bank

Challenge Activity
possible answer—Israelis and Arabs can share the land and make Jerusalem a shared capital. Both governments can rule over the land. This proposal would be successful because it is fair to both sides.

Teacher Management System

SECTION 4
Call-Out Boxes
1. France
2. Syria's government owns the country's oil refineries, larger electrical plants, railroads, and some factories.
3. in the 1940s
4. Muslim and Christian
5. after World War I
6. In addition, the tourism and banking industries are growing. Jordan depends on economic aid from oil-rich Arab nations and the United States.

Challenge Activity
possible answers: Syria—share its wealth with the people more; Lebanon—agree to disagree about religion and stop fighting with each other; Jordan—expand its tourism industry by creating a safe environment for tourists

The Arabian Peninsula, Iraq, and Iran

COMPREHENSION AND CRITICAL THINKING
1. Islamic religion and culture, monarchy as a form of government, and valuable oil resources
2. It has changed from having a king, or shah, to being an Islamic republic which follows strict Islamic law. Today, Iran is a theocracy ruled by religious leaders that restrict the rights of most Iranians.

SECTION 1
Call-Out Boxes
1. sand, bare rock, gravel
2. rivers, plains, plateaus, mountains
3. No, it can get very cold at night.
4. Their roots either go very deep or spread out very far to get as much water as they can.
5. water and oil

Challenge Activity
Students' posters will vary but should contain words accurately describing every letter in the term *Persian Gulf*.

SECTION 2
Call-Out Boxes
1. Muhammad in Saudi Arabia
2. It has very little freshwater to grow crops, so it has to import most of its food. There is high unemployment.
3. Kuwait, Bahrain, Qatar, the United Arab Emirates, Oman, and Yemen
4. banking, tourism, and natural gas

Challenge Activity
Answers will vary but should be based on the physical and human geography of the region and include reasons for why the strategies might work.

SECTION 3
Call-Out Boxes
1. in Iraq, in the area called Mesopotamia
2. He restricted the press and people's freedoms and killed political enemies.
3. embargo
4. The U.S. government believed Iraq aided terrorists in the September 11, 2001 terrorist attacks.
5. Arab
6. The population of Baghdad is more than twice the size of Los Angeles' population.
7. to write Iraq's new constitution

Challenge Activity
Outlines will vary but should address important government issues.

SECTION 4
Call-Out Boxes
1. It was important to them and was a part of their heritage.
2. follows strict Islamic law
3. more than 50 percent
4. People celebrate the Persian New Year, Nowruz, and eat Persian food at family gatherings.

Teacher Management System

Challenge Activity

Diary entries will vary but should reflect what life might have been like at the time the Muslims conquered the Persian Empire. They may include feelings of sadness or anger about what was happening.

Central Asia

COMPREHENSION AND CRITICAL THINKING

1. oil and gas, minerals, and water
2. rugged mountains, landlocked
3. gold—minerals, shrinking Aral Sea—environmental issues, yurts—ethnic influences

SECTION 1
Call-Out Boxes

1. landlocked, rugged land
2. Two important rivers are the Syr Darya and the Amu Darya.
3. Aral Sea
4. Kara-Kum, Kyzyl Kum
5. water, oil and gas, minerals such as gold, lead, and copper

Challenge Activity

possible key facts—harsh, dry climate with extreme temperatures; physical features include mountains, plains, lakes, rivers, and deserts; prone to earthquakes; resources include oil, gold, lead, copper; possible list of supplies—food and water; coats and blankets for the cold, umbrellas for shade from the sun; veil to keep blowing sand out of face and eyes

SECTION 2
Call-Out Boxes

1. Europe and India, Europe and China
2. They discovered they could sail to East Asia on the Indian Ocean.
3. from the 700s to 1200s
4. They became independent countries.
5. yurt
6. Cyrillic

Challenge Activity

Students' travel journal entries might describe many different kinds of products and wares being transported, and people from many different cultures traveling together and interacting with one another.

SECTION 3
Call-Out Boxes

1. the Taliban
2. But it is growing again, because of oil reserves and a free market.
3. Only 5 or 6 percent of the land is arable.
4. shrinking Aral Sea, leftover radiation, crop chemicals harming farmlands

Challenge Activity

Students' posters might appeal to people's sympathies, showing people who are sick or devastated farmlands. They might also show how money can be used to help restore the environment.